WILDLIFE

PONDS AND STREAMS

Ian Russell

David & Charles
Newton Abbot   London   North Pomfret (Vt)

British Library Cataloguing in Publication Data

Russell, Ian
　Ponds and Streams. – (Wildlife)
　1. Pond fauna – Great Britain
　2. Pond flora – Great Britain
　3. Stream fauna – Great Britain
　4. Stream flora – Great Britain
　I. Title　　II. Series
　574.92' 9' 41　　QH137

ISBN 0-7153-8162-8

© Ian Russell 1982

All rights reserved. No part of this
publication may be reproduced, stored
in a retrieval system, or transmitted,
in any form or by any means, electronic,
mechanical, photocopying, recording or
otherwise, without the prior permission
of David & Charles (Publishers) Limited

Typeset by Typesetters (Birmingham) Limited
and printed in Great Britain
by A. Wheaton & Co, Exeter
for David & Charles (Publishers) Limited
Brunel House　Newton Abbot　Devon

Published in the United States of America
by David & Charles Inc
North Pomfret　Vermont 05053　USA

Contents

1	**Freshwater life**	4
2	**How to collect pond and stream animals**	8
3	**Water plants**	
	Common reed, reedmace, bulrush, waterlilies, starwort, pondweed, duckweed, frogbit, water soldier, hornworts, water milfoils, Canadian pondweed, algae	11
4	**Among the water plants**	
	Snails, water beetles, damselflies, mayflies, caddis fly, water scorpion, water boatmen, water mites, water louse, freshwater shrimps, flatworms, leeches, hydras	14
5	**Among the stones of fast streams**	
	Water moss, fast-water mayfly larvae, stonefly larvae, freshwater shrimps, snails, river limpet, flatworms, leeches, caddis larvae, blackflies	19
6	**On the water surface**	
	Pond skaters, water crickets, water measurer, whirligig beetles, mosquito larvae	23
7	**In the mud**	
	Algae, midge larvae, segmented worms, freshwater mussels and cockles, mayfly larvae, dragonfly larvae, alder fly larvae	30
8	**Very small animals and plants**	
	Filamentous algae, flagellates, ciliates, amoeba, rotifers, roundworms	33
9	**Larger animals**	
	Water vole, water shrew, otter, moorhen, duck, great crested grebe, grey wagtail, dipper, kingfisher, frogs, toads, newts, stickleback, minnow, eels, trout, loach, bullhead, roach, perch	38
10	**Aquariums and freshwater communities**	42
11	**Collecting with a purpose**	47
12	**Pollution**	50

1
Freshwater life

Children have always been fascinated by the teeming underwater life of ponds, ditches and streams. Any summer's day you can watch the jamjar brigade poking with nets in the murky water of the local park. A survey at a symposium at Oxford once indicated that a large percentage of eminent zoologists could trace the origin of their careers to an early interest in pond life.

Junior pondhunting still thrives on a vast, international scale; is it the stirring of a primitive hunting instinct or some mysterious memory of our aquatic ancestors?

One of the attractions is that the world of freshwater animals is so accessible. The many forms of pond and stream life are amazingly widespread, so that in the country you need look no further than the nearest weed-choked pool or water-filled ditch. Even in towns many parks have boating lakes or some sort of ornamental pool or watercourse that is full of underwater wildlife. Very few homes are far away from ideal freshwater habitats, although the 'typical' countryside pond that we all imagine is fast disappearing. We visualise a rustic pond in the corner of a field, shaded by trees and with part of the margin conveniently freed from choking vegetation by the trampling of drinking cattle: farmers get government grants for filling in such ponds and many others have vanished beneath new housing estates.

In some unfortunate areas there may be little sign of life in water contaminated by the chemical waste of factories, or in streams suffocated by decaying sewage or agricultural waste like silage. The very justifiable fuss being made about pollution must not blind us to the fact that many freshwater habitats are alive and well in most areas. We are often too ready to write them off as polluted. For example, the word 'stagnant' has unpleasant overtones, yet there is no harm or health risk in stagnant water in this country, unless you drink it. Whatever your feelings about the water beneath an unbroken scum of duckweed, it is a natural, healthy, balanced environment and harmful substances will only be present if they have been introduced by man. Similarly, very few streams contain sewage. If you find one that does, inform the local Water Authority.

This book has been entitled *Ponds and Streams* because these are so much more numerous and accessible than lakes and rivers. Even so, much of what is said here is also applicable to the shallow margins of lakes and rivers, provided that certain differences are borne in mind. Anything more than a stream or a small pond is likely to contain much larger fish than the familiar minnows and sticklebacks. Many lakes are so deep that insufficient light reaches the bottom to support plants, so their underwater scenery consists of dim plains of flat mud. Countless species live exclusively in this mysterious domain and many new ones undoubtedly await discovery, (possibly including the Loch Ness Monster).

Another difference with these larger bodies of water will be noticed if you poke about in the rocky margin of a lake. The creatures you find there bear more resemblance to the inhabitants of a fast stream than to those of a pond, many of them clinging tightly to the stones. This is because large waves frequently cause violent water movements at the edge of a lake.

The smaller bodies of fresh water, the ponds, streams and ditches of our land, can be explored by anybody almost anywhere. There is no need to travel far, no costly equipment is necessary, and all sorts of fascinating organisms are easy to catch and study. It is a world about which a tremendous amount remains to be discovered, though several research establishments are devoted entirely to it.

2
How to collect pond and stream animals

A large selection of pond and stream animals can be found simply by grubbing around with your bare hands at the water's edge, but collecting is far more productive with suitable equipment. A useful and professional-looking collecting kit can easily be put together.

Top of the list comes some sort of shallow, watertight container in which to sort specimens from mud and weed. It must provide a pale-coloured background to help you spot small creatures in a centimetre or so of muddy water. The bigger the better, but it has to be small enough to carry about. The shallow type of white plastic ice-cream box is just about big enough; a white plastic tray as supplied for photographic darkrooms is even better.

A teaspoon and a cheap little paintbrush are useful for handling specimens. Try to get hold of a couple of pipettes from a chemist's shop — short glass tubes with a small rubber teat at one end and a narrow 'nozzle' at the other. Two pipettes are best, one with a fine nozzle for very small animals and the other without. A nozzle can be reversed and stuck inside the teat, or it may be carefully snapped off and the sharp glass rounded off by heating in a gas flame.

A magnifying glass will often be useful. It need not be particularly powerful: ×5 is strong enough. Strong lenses need to be held very close to the specimen, which can be awkward if it is under water. A lens which folds away into a protective cover is a good idea.

A glass jam jar makes a convenient waterside aquarium in which to study individual creatures, with or without a magnifying glass. A couple of screw-topped jam jars are also useful for carrying specimens.

The best way to take home a wide variety of smaller specimens is to obtain a dozen or so proper specimen tubes. These are like small bottles of glass or clear plastic which do not become narrower at the top and have tight-fitting stoppers or screw-on caps. If you can't find a dealer in laboratory equipment, try a chemist or visit a hypochondriac relative, because many plastic pill-containers are suitable.

A handful of smallish polythene bags with some wire fasteners will

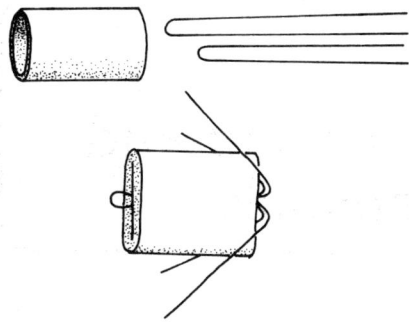

Making a weed drag

enable plants to be taken home for closer examination or for an aquarium. An inexpensive cloth bag with a shoulder strap is necessary to carry this lot.

A weed drag is another indispensible piece of equipment (see diagram). Bend in half two 8 inch lengths cut from a wire coathanger and push them into a 2 inch length of lead or copper pipe. Bash the pipe completely flat with a heavy hammer and bend the four prongs into position. If you leave a loop of wire projecting from the other end of the flattened tube, a length of strong cord can be tied to it, then wound on to a square piece of wood. Something nearly identical costs several pounds from a supplier of biological equipment.

The drag is used to capture the many organisms that inhabit weedbeds. First, half-fill the plastic ice-cream box with pond water, then unwind a suitable length of cord and heap it neatly on the ground. Holding one end of the cord, throw the weed drag into the weedbed you wish to sample. If all goes well you should be able to haul in a fair-sized lump of weed. Pull it in as fast as possible, before too many of the more active animals have a chance to escape. Lift out the mass of weed and dump as much of it as you can into the box of water. Unless you particularly want to examine the various inhabitants of mud, try to keep the plant roots out because they cloud the water. Shake the weed vigorously for a while, then lay it aside. With any luck your catch will contain a surprisingly wide variety of animals and you can sit down to sort them out. Some likely finds are described in chapter 4.

This is the simplest way to catch small freshwater creatures in ponds, ditches and slow streams. By moving around it is interesting to compare the animals inhabiting different species of weedbed living at different depths. This method is only likely to fail to produce impressive results in winter, when many water plants die back and the populations of small animals are greatly reduced.

Various species of minute animals spend most of their time swimming and drifting mid-water, often well

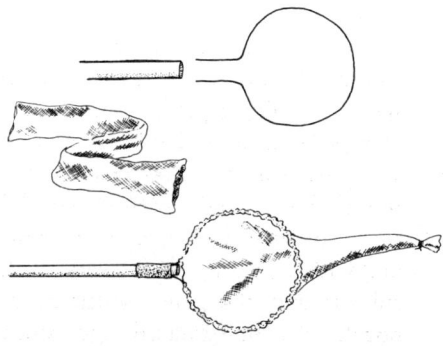

Making a plankton net

7

away from weedbeds. They are collectively known as 'plankton', a term also applied to the minute drifting creatures of the open sea. A very effective plankton net can be made from the leg of a pair of nylon tights (see diagram). Bend a loop about 5 inches across from some more coathanger wire or something similar. Bind it securely to a 4 foot bamboo cane, using many tight turns of plastic insulation tape. Cut an 18 inch length from one leg of the tights and stitch it to the wire frame. Close the other end of the net, tying it off with thread. Do not make the net any shorter than specified or water will not flow fast enough through the mesh, producing an underwater 'bow wave'.

To sample an area of water, the plankton net is moved continuously to and fro for a while, in a figure-of-eight pattern. The knack is quickly acquired with practice. The catch is removed by turning the net inside out and immersing the end in a couple of inches of water in a jam jar or the sorting tray. The abundance of freshwater plankton varies a lot from place to place, even in the same pond, but the net will sometimes come out bulging. The several species of small crustaceans called 'water fleas' (see chapter 4) usually make up the bulk of the catch, together with a selection of small insect larvae and a few beetles.

This net can also be used to catch larger individual animals, including insects on the surface film, and to dredge samples from the surface of the bottom mud.

Here is a simple checklist of pond-hunters' equipment, capable of sampling most forms of freshwater life.

Sorting tray (ice-cream box or tub); teaspoon; paintbrush; narrow-mouthed and wide-mouthed pipette; magnifying lens; two screw-topped jam jars; specimen tubes or plastic pill-boxes; polythene bags; bag; weed drag; plankton net.

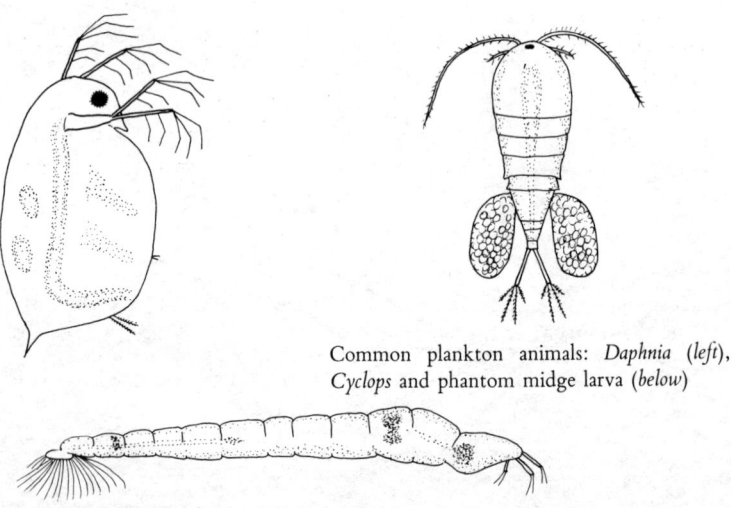

Common plankton animals: *Daphnia* (*left*), *Cyclops* and phantom midge larva (*below*)

3
Water plants

There are two main categories of water plants. The simplest ones are the microscopic algae that sometimes form slimy green masses, or give a green tinge to mud or even to the water itself. The algae are more fully described in a later section. Most of the remaining water plants belong to the same group as the land plants, with stems, roots, leaves and flowers. In the summer, ponds, streams and ditches usually support a rampant confusion of plant life along their margins and beneath the water's surface. Quite apart from its own interest, the luxuriant growth should be investigated because of the tremendous amount of small animal life that shelters in it.

Even some distance back from the water's edge the soil is permanently moist. Low, green clumps of rushes are everywhere, together with various colourful flowers. The elegant yellow flag iris is one of the brightest of these.

The plants right at the water's edge are content to grow 'with their feet in the water'. Most are tall and well rooted in the mud. The common reed is very widespread, pencil-thin yet taller than a man and with its feathery flowers swaying gracefully in the breeze. The bulrush (*Schoenoplectus lacustris*) is also common here but presents us with a naming problem. The familiar, brown-clubbed plant

Common reed (*Phragmites communis*) (*left*) and reedmace (*Typha latifolia*)

also seen in this habitat, and always shown in old pictures of the infant Moses, is *not* the bulrush, insist the botanists: it is the reedmace (*Typha latifolia*).

Other plants grow underwater in the shallows, yet tend to form masses of floating leaves and flowers. The water lilies are the largest, with almost circular leaves and big beautiful waxy flowers. There are two common species, the white water lily (*Nymphaea alba*) and the yellow (*Nuphar lutea*).

Tangled masses of starwort (*Callitriche verna*) are often topped with floating leaves in shallow water. It is a very common plant with a long, slender stem and leaves in opposite pairs, forming star-shaped rosettes at the surface. The broad-leaved pondweed (*Potamogeton natans*) has larger, oval, floating leaves. In some places the surface may be covered with the floating leaves and white flowers of various species of water crowfoot. These plants, along with many other water plants, produce two distinct shapes of leaf. There is a rounded floating type and a finely divided submerged version which gives the plant its name.

Water plants obtain a large proportion of the materials they need directly from the water. Their roots are not as important as those of land plants. Some species are therefore able to live afloat on the water surface with no firm anchorage. The little duckweed plants are familiar examples, often forming an unbroken green film over the surface. A root hangs down from

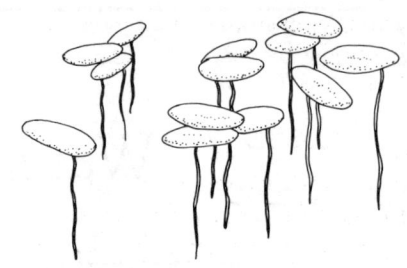

Duckweed (*above*)

Water milfoil (*Myriophyllum* spp) (*top*) and hornwort (*Ceratophyllum* spp)

each tiny plant, absorbing essential dissolved chemicals from the water. There are also larger floating plants such as the frogbit (*Hydrocharis morsus-ranae*), with a rosette of circular leaves, and water soldier (*Stratiotes aloides*), with long, spiky leaves.

The hornworts are fully submerged plants which lack roots and have straight, slender stems and whorls of narrow leaves. They often float freely in a vertical position.

There are many species of fully submerged plants which form dense beds deeply rooted in the bottom mud. The water milfoils are attractive, with reddish stems and feathery leaves. They make good aquarium plants. So does the Canadian pondweed (*Elodea canadensis*). This very common plant looks rather like the starwort, but is darker green and more brittle. Although common now, this plant was unknown in Britain at the beginning of the previous century.

Where submerged water plants are absent from unpolluted water, it is usually because of lack of light. They will not receive enough light if the water is too deep, or if mud is stirred up constantly, perhaps by cattle or boats. In other places a surface layer of duckweed may critically reduce light penetration.

The ability of a body of water to support plant growth is closely related to the amount of animal life it can feed. It has been mentioned that water plants are an important form of shelter

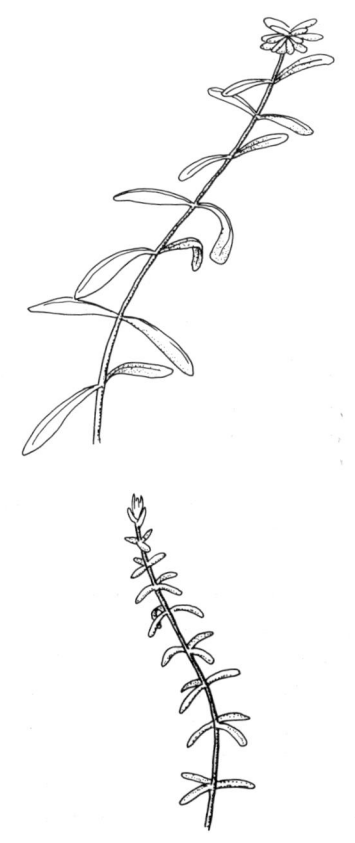

Starwort (*Callitriche* spp) (*top*) and Canadian pondweed (*Elodea canadensis*)

for small animals. More than this, those humble microscopic plants — the algae — are the basic source of food for the entire freshwater community. These algae are the freshwater equivalent of the grass on the African plains which effectively supports everything from antelope to lions and vultures.

4
Among the water plants

A dense jungle of underwater plants attracts small animals: it provides food for some of them, and most find shelter from their enemies and a convenient surface to hang on to.

Imagining now that a handful of weed has been shaken in a little water in a shallow white container, then removed, there follows a general description of the commonest small animals likely to be found as you search about in the slightly muddy water. This technique is usually surprisingly successful, but not *all* these creatures can be expected in a single sample.

Water snails are usually abundant on water plants. There are a great many different species and many of them grow no larger than a fraction of a centimetre. The wandering snail (*Lymnaea peregra*) is the commonest, with a broad-mouthed shell rather less than 2 cm long. The various species of ramshorn snails will be familiar to aquarium owners. These have shells that coil in a flat spiral. The biggest snail is the great pond snail (*Lymnaea stagnalis*) which grows to over 5 cm in length. Some snails will eat water plants, but their main food consists of the coating of microscopic green algae. The rasping mouth of a snail can be seen in action against the glass of a jam jar or aquarium. Snail eggs often appear on the glass of an aquarium in the form of masses of clear jelly full of tiny specks of life.

The active little water beetles are immediately noticeable. Most aquatic insects are just young larval forms that develop into flying adults, but water beetles spend much of their time under water, even though they are capable of flying. There is a huge number of different species and some are carnivorous, others vegetarians. The biggest is the great diving beetle (*Dytiscus marginalis*) which is over 3 cm long. Many others are very small.

Water beetles breathe air from the surface and a silvery bubble of air is often visible, trapped against the abdomen, as the beetle struggles against the extra buoyancy of this reserve supply.

The young larvae of water beetles are variable in shape, but are basically elongated, segmented and with shortish legs. These larvae normally crawl out of the water before emerging from their skins as adult water beetles. (This should be borne in mind if this transformation is to be followed in an aquarium.)

The damselflies are one of the many insect groups which spend their early life underwater, feeding and growing before changing into non-aquatic winged adults. Damselfly larvae are fairly common amongst water plants, where they live as carnivores preying on all sorts of smaller animals. They can be recognised by the three leaf-shaped gills at the rear of the long abdomen. Eventually, the fully grown larva crawls out of the water and sheds its dowdy skin to reveal an adult insect of glittering beauty. There are a number of species, but most are highly colourful with narrow, metallic-coloured bodies a couple of inches long. Some have transparent wings which become invisible in flight. Others have coloured wings and flutter like butterflies. In the summer it is common to see mating

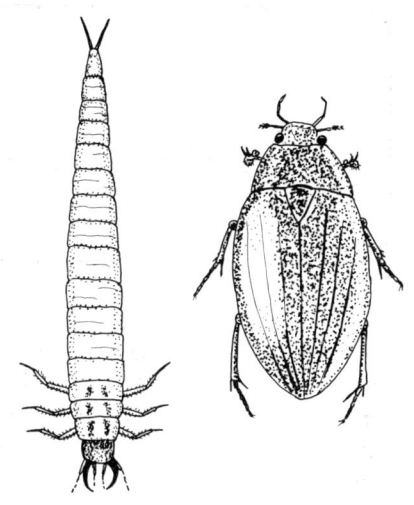

Water beetle: larva (*left*) and adult

pairs flying joined together.

Damselflies attack smaller insects resting on the waterside vegetation. They are close relatives of the much larger, fiercer dragonflies whose larvae live in the mud at the bottom of the

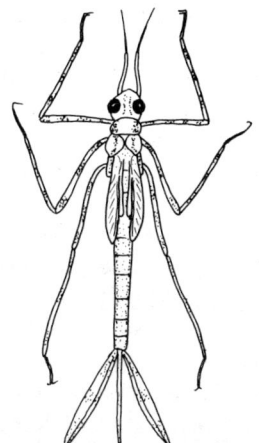
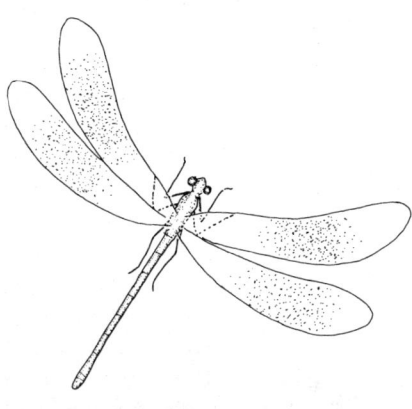

Damselfly: larva (*left*) and adult

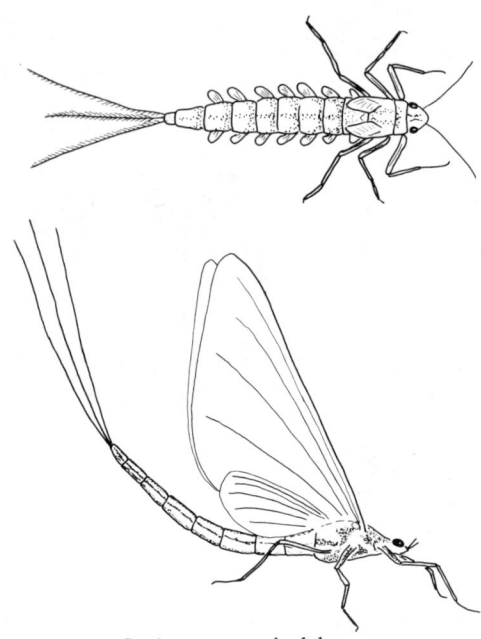

Mayfly: larva (*top*) and adult

The larvae of the moth-like caddis fly live in all sorts of freshwater environments. There are many different species. Most are characterised by a tubular 'shell' made from stones, sandgrains, twigs or pieces of water plant. The larva drags this around and retreats into it when danger threatens.

The sluggish, but fearsome-looking water scorpion is about 3 cm long and as carnivorous as it looks. It resembles a dead leaf when it lies in ambush in shallow water. The long breathing tube at its rear end is used to supply air from the surface.

There seems to be widespread confusion between the water boatman and insects like the pond skater which live on the surface film. The various species of water boatmen live underwater. Propelled by their oar-like legs, they are more like 'rowing submarines'. They are extremely common and there are two main groups. The true water boatmen swim on their backs with a silvery film of air covering their belly. They are carnivorous, and the piercing mouthparts of the larger species are quite capable of pricking a finger. The other group, the lesser water boatmen, swim the right way up and feed on the surface of the mud, sucking up the organic matter.

Sometimes a tiny bright-red speck will be seen swimming around with a wandering, wavering motion. This is a water mite. Not all species are red, others are bright green or pale brown. Unlike the other creatures described, these are not insects. Close examination reveals four pairs of legs, indicat-

pond, and are described later.

The largest species of mayfly have bodies up to 2 cm long and large wings and tails, but these are only locally common. There are numerous smaller species to be found almost everywhere. Both adults and larvae can be recognised by the characteristic group of three long tail appendages. (Mayfly larvae can be distinguished from damselfly larvae by the length and thinness of these tail appendages, and by the presence of gills along the sides of the abdomen.)

Although the mayfly larva may live and grow for a year or so, the story that the adults often lay their eggs and die within twenty-four hours of emerging from their larval skin is perfectly true. Their up-and-down mating flights can often be seen.

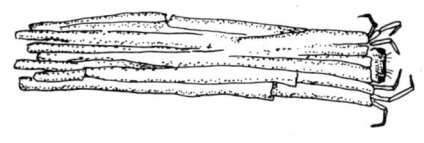

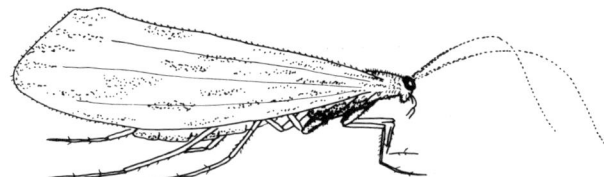

Caddis fly: larva in case (*top*) and adult

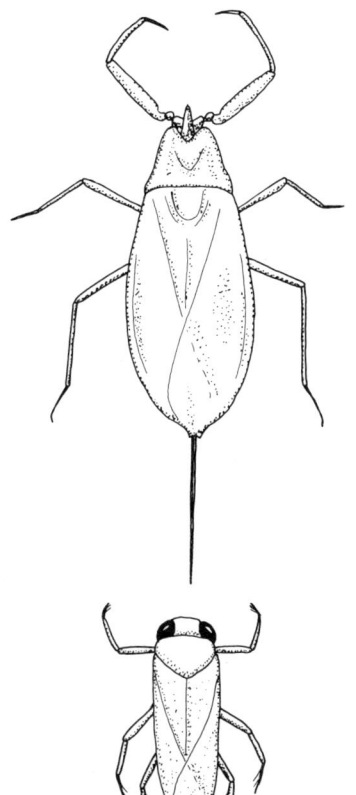

Water Scorpion (*Nepa cinerea*) (*top*) and lesser water boatman (*Corixa* spp)

ing that they belong to the same group as the spiders, the Arachnids. They are carnivorous, devouring any minute animals they encounter.

Water lice and freshwater shrimps are not insects either but crustaceans (see page 20). The water louse resembles its more heavily built relative, the woodlouse. The freshwater shrimp also has too many legs to be an insect, and is flattened from side to side, unlike the water louse which is flattened from top to bottom. It is not a true shrimp, being more closely related to the sandhoppers which swarm in rotting seaweed on the seashore. Water lice and freshwater shrimps are both common in weedbeds. They are general scavengers.

Flatworms have the appearance of brown or grey oblong smudges, a centimetre or so in length (see page 44). They glide smoothly and quite rapidly along the bottom of the sorting tray. The mouth is midway along the undersurface and they are mainly carnivorous. It is easy to capture them by setting a bait, such as a piece of meat tied to a string, leaving it for a while, then pulling it out and washing off the flatworms.

The various species of leech look vaguely similar to flatworms until they start to move. They get about by means of suckers at the front and back end, with highly characteristic 'looping' movements. All are carnivorous, and many species suck the blood of larger animals. Only the medicinal leech can pierce human skin. This species is only found in a few areas and is easily recognised by its large size (over 10 cm). Some leeches may be seen swimming with vigorous up-and-down undulations of their bodies.

Hydras are minute animals related to the jellyfish and anemones of the seashore. There are several species, but they are not likely to be found during routine collecting, because they are hard to see.

Hydras feed on small creatures such as water fleas and they are ideal specimens for the 'window-ledge zoo' and for observation with a microscope.

The best way to collect them is to place various samples of water plants in a number of jam jars of pond water and leave for a few days. Hydras may often then be seen on the sides of some of the jars. Their thin bodies are usually about 1 cm long when extended and very thin tentacles hang down from the end. The green hydra, is attracted to light, so partial shading of the jars can make them move.

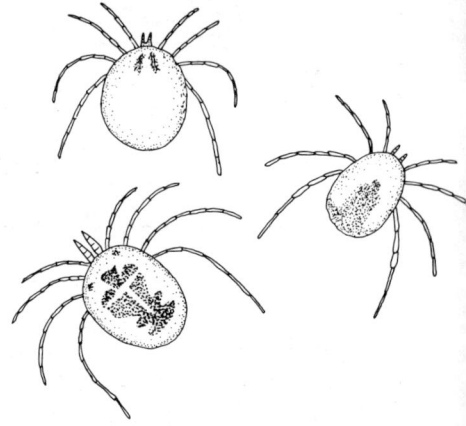

Water mites

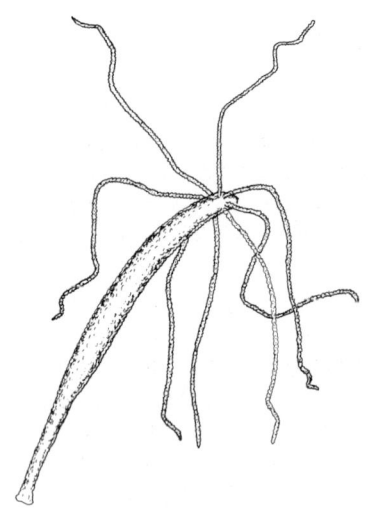

Hydra

5
Amongst the stones of fast streams

There is no hard and fast division between the creatures and plants of still-water habitats such as ponds and ditches, and those of the running water of rivers and streams. Many species are found in both. The previous section described the animals likely to be encountered while searching amongst the water plants of ponds. Nearly all these will also be found in the shelter of weedbeds in slow flowing streams. There is little difference as far as many animals are concerned, although certain species do show a marked preference for either still or flowing water.

Nevertheless there are considerable differences between the conditions in a pond and those in a fast flowing stream. As we look at the characteristics and inhabitants of a fast stream, we have to bear in mind that at various times we are likely to find every intermediate stage between the two extremes represented by fast flowing and static water.

First, a current washes away soil and mud, leaving a bed of stones or gravel. Water plants are only able to grow in areas sheltered from the main current, but dark green growths of water moss cover many of the larger stones. A strong current presents problems to animals as well. Only fish, such as trout, minnows, loach and bullheads have sufficient strength to swim against it. The smaller creatures have only two choices if they are not to be swept away helplessly. They can cling tightly to the stones, or shelter below them from the main force of current. Many do both. Many of the mayfly larvae found here are extremely flattened, of a shape allowing the current to press them tightly against the stones.

There are two compensating benefits for the inhabitants of fast streams. For one thing, there is no need to

Water moss (*Fontinalis* spp)

venture far in search of food. Decaying plant fragments, algae, micro-organisms and small, dislodged creatures are swept constantly past. A number of specialised fast-water animals build nets and traps to make use of this free food supply.

The other advantage is the large amount of dissolved oxygen in the water. This gas, essential to animals, can often become scarce in stagnant water. But here the splashing and turbulence constantly replenish the supply. Many fast-water species are unable to live for long in jam jars or aquaria without strong aeration.

We can easily find out what animals live here by looking beneath stones, carefully replacing them afterwards. As a stone is lifted, many specimens remain clinging tightly to its lower surface. Others may be swept away downstream into a suitably positioned hand net.

When a stone is lifted from the stream bed the attention is often caught by the wriggling movements of small mayfly larvae in the film of water beneath it. There are many species and they are not likely to be the same as those found in still water, although they share the same characteristics: three long tails, and gill-fringed bodies. The fact that many remain on the stone when it is pulled out of the stream demonstrates how well they can cling to it to resist the current. The flattened shape of many fast-water mayfly larvae is ideal for their clinging existence.

Stonefly larvae are common in running water and resemble mayfly

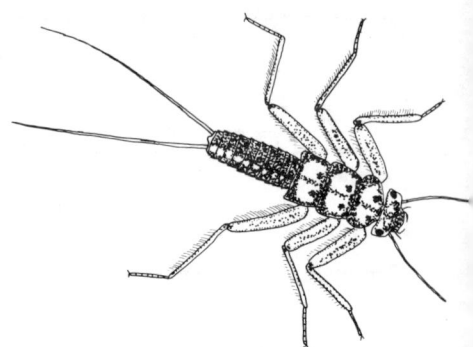

Stonefly larva

larvae at first glance. However, they only have two tails and lack the well developed gills of the mayflies. Their rudimentary gills restrict them to habitats with high oxygen levels: they are not found in ponds and are hard to keep in captivity. Certain species are found amongst the stones on the wave-washed shores of lakes, a habitat similar in many ways to fast streams. (There are many other similarities between the populations of certain lake margins and rocky streams.) As with mayflies, identifying the various species of stoneflies is a task for the specialists.

Freshwater shrimps will also often be noticed on the undersides of stones, skipping about on their sides in their familiar way. There are only a couple of species and the ones you find are probably of the same species, (*Gammarus pulex*) that is most common in still water.

Tiny, dark, elongated snails less than half a centimetre long are often common. This is Jenkins' spire shell, (*Potamopyrgus jenkinsi*) which represents something of a snail success story. At the end of last century this snail was unknown in fresh water, and lived only in the brackish water of estuaries. For some unknown reason it has suddenly invaded rivers and streams and is now just about the commonest snail in running water. Various other snails will also be seen, mostly small ones. Among them are various species of ramshorn snails with their flattened, 'catherine wheel' shells.

Better adapted for clinging to stones in very fast currents is the river limpet (*Ancylastrum fluviatile*) with its small conical shell, well under a centimetre long. A related species is found in the margins of lakes.

Less likely to catch the eye when examining a stone are the soft, flattened bodies of flatworms and leeches, but several species of each are common. Although similar in outline, flatworms and leeches are different animals, and can be instantly distinguished as soon as they move. A leech will usually search from side to side with its pointed front end, while its broader rear remains firmly attached to the stone. Finally it will take hold with the front sucker and move off with its unmistakable 'looping' action. The various flatworm species all glide along smoothly rather like snails.

Little clusters, a centimetre or two long, formed by numbers of tiny pebbles, are often seen on stones from a stream bed. They are the homes of highly specialised caddis larvae which have given up any idea of carrying their homes around and fasten themselves down in a fixed position instead. They then rely on the current to bring them an adequate supply of food. As we saw, the various species of caddis larvae found in still-water weedbeds normally occupy tubular mobile homes made of pieces of water plants. Mobile caddis larvae are also found in quite fast streams, but their 'homes' are more often heavier affairs made of small pebbles. It is not just a question of the shortage of water plants: heavy structures are a definite advantage where there is any risk of being swept away downstream.

Comparing the inhabitants of ponds and streams like this can illustrate how all living creatures have adapted to the particular ways of life they lead.

The various species of blackfly have a larval stage which is particularly well adapted for life in fast water. Large numbers of the larvae, which

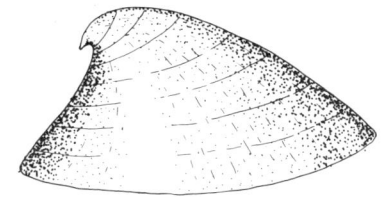

River limpet (*Ancylastrum fluviatile*)

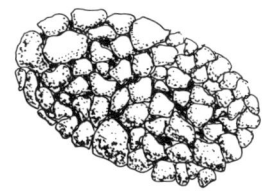

Case of a fast-water caddis larva fastened to a stone

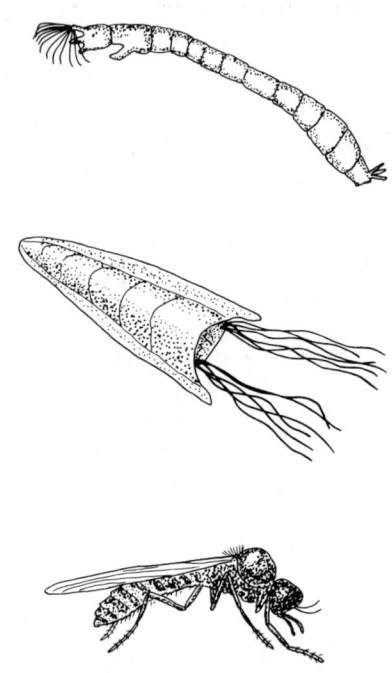

Blackfly (*Simulium* spp): larva (*top*), pupa and adult (*below*)

look rather like tiny caterpillars just over a centimetre long, may be seen clinging to stones, branches and water plants, often where the current is strongest. They sieve out their food from the water with comb-like bristles at the front end, while firmly anchored down by the rear end. If dislodged, they move with a looping action that heightens their resemblance to certain caterpillars.

When the larvae have become pupae, the waiting stage before the emergence of the winged adults, they look totally different, like dark little cones cemented firmly against any solid object. The wide end of each cone always faces into the current. When the adult fly emerges from the pupal skin, it has no time to waste before it escapes from the swirling torrent at the surface. It bursts free and bobs to the surface surrounded by a bubble of air. Popping out through the surface film, it flies away instantly.

This insect, again, is wonderfully adapted to its environment. The adult fly, however, has a fearful reputation as a biter. I had an unpleasant encounter with a particularly nasty species of blackfly once while clipping a hedge in the North of Scotland. Large numbers must have been resting in the hedge, because they emerged from the clippings on the ground and went up the legs of my jeans. Surprisingly I felt nothing during the attack, presumably because of the anaesthetic injected by the mouthparts of many bloodsuckers. A chemical which prevents clotting and keeps the blood flowing is also injected, and the first I knew about it was that I noticed that my socks were soaked with blood. Marks were still visible on my legs six months later, and I have had a healthy respect for blackflies ever since.

6
On the water surface

Water has certain peculiar properties which we, as large animals, seldom notice. Small pond creatures experience it as a very different substance. A bather can swim to the edge of a swimming pool and hoist himself out casually. Think about that next time you help a drowning fly out of a pool of water. If you just push the fly to the edge, it will often be unable to drag itself out. The water clings to it and seems to be dragging it back. The surface curves and is pulled up as the insect heaves — and insects, for their size, are the strongest creatures in the world. It is as if the water has a surface skin which will not break and will not release the waterlogged creature.

This is the phenomenon of 'surface tension', a mere curiosity to us, but a ruling factor in the lives of many small water creatures. It has its uses as well as its dangers for some of them. Certain insects can walk on the surface film without breaking through it and getting wet. Just try doing *that* at the swimming baths. The secret, apart from being of small size, is to have a waxy covering on the body to which water will not stick, and to have pads of similarly waxy bristles on the feet to avoid breaking through the surface. Insect species equipped to 'walk on the water' in this way can be seen everywhere in the summer, in ditches, the margins of ponds and the sheltered areas of slow flowing water at the edges of streams.

However, most insects do not have this ability to anything like the same degree. For them, the surface of the water can be a death-trap because once they break through that strange 'skin' it holds them and drags them down. Small insects are so abundant at the waterside that the surface is often covered with struggling or dead individuals who have become trapped. These unfortunate insects are the main food of the water-walkers. Their drowning struggles send out widening circles of tiny ripples. The carnivores of the surface film are highly sensitive to these and know instantly where any surface disturbance is coming from. In this way they locate their prey.

The pond skaters are perhaps the commonest creatures of this type. There are a number of species and their long legs make them look quite

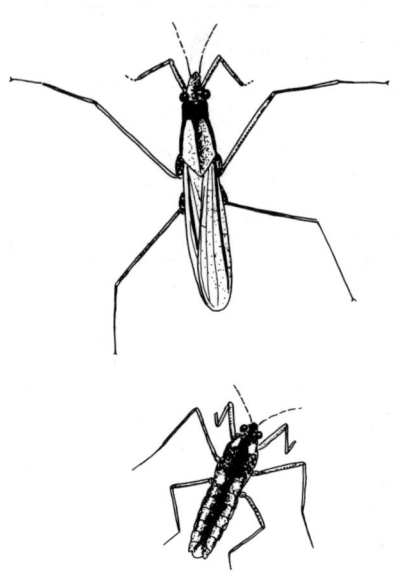

Pond skater (*Gerris* spp) (*top*) and water cricket (*Velia* spp); (*below*) water measurer (*Hydrometra stagnorum*)

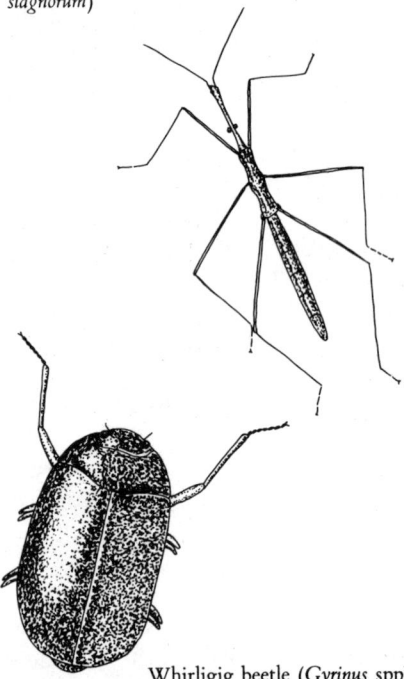

Whirligig beetle (*Gyrinus* spp)

big. All insects, of course, have six legs yet at first glance the pond skaters seem only to have four. A closer look shows that the front pair of legs are short and powerful, held at the ready in front for grabbing hold of struggling prey.

The water crickets, dark brown with orange markings, are commonest on running water.

The water measurer is just over a centimetre long and extremely thin. Its name comes from the slow, deliberate way it paces across the surface. It is less commonly seen than the skaters and water crickets.

The three groups mentioned above are all 'bugs'. The characteristic features of this group are the piercing, tubular mouthparts designed to suck out the soft insides of their prey. The whirligig beetles belong to an entirely different order of insects, with their typical beetle-type hard wing-cases.

Whirligig beetles are incredibly active. They first resemble swarms of small, shiny ball-bearings skimming over the surface in mazy zig-zags. When you stoop for a closer look they dive beneath the surface and vanish. They feed on dead and drowning insects, but their 'ripple-sense' has an added refinement. Their swimming activity sends out minute, buzzing ripples, and they can detect and decode the patterns made when these ripples are reflected back from objects on the surface. It is a kind of 'ripple radar', which explains how they can locate dead as well as struggling insects. It also explains how these astonishing insects avoid bumping

Freshwater algae under the microscope

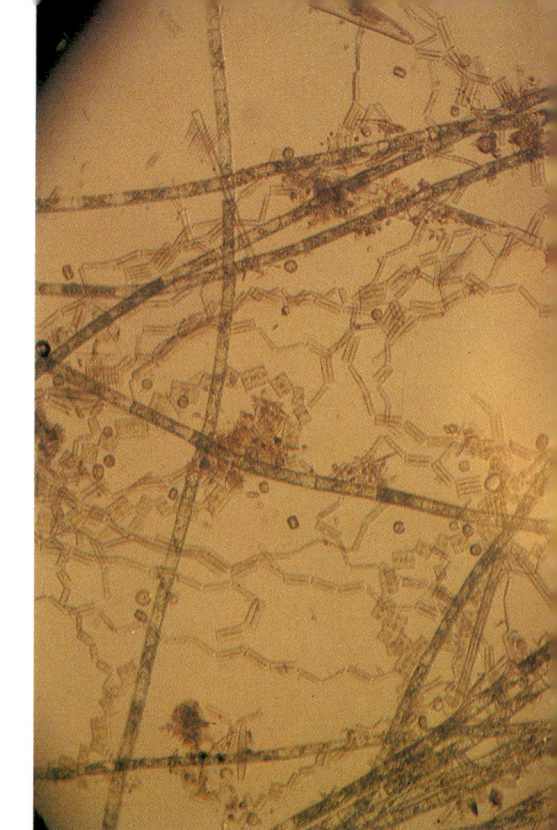

Fenland pond with a clump of purple loosestrife, and the leathery, heart-shaped leaves of the yellow water lily on the pond surface (*Leslie Jackman/Wildlife Picture Agency*)

(*Above*) Young spring growth at the edge of a freshwater marsh, showing ferns and flag spikes appearing (*Rodger Jackman/Wildlife Picture Agency*)

(*Opposite*) Fast flowing stream in a field, bordered by the yellow flag, wild relative of familiar garden irises (*John Beach/Wildlife Picture Agency*)

(*Below*) Water ditch showing reddish deposits from the oxidisation of iron salts (from surrounding soil) by bacteria. This indicates low oxygen levels, unsuitable for much animal life, but the process encourages the growth of algae (*David Cayless/Wildlife Picture Agency*)

Pond with floating leaves of white water lily, marginal growth of common reed and domesticated muscovy ducks

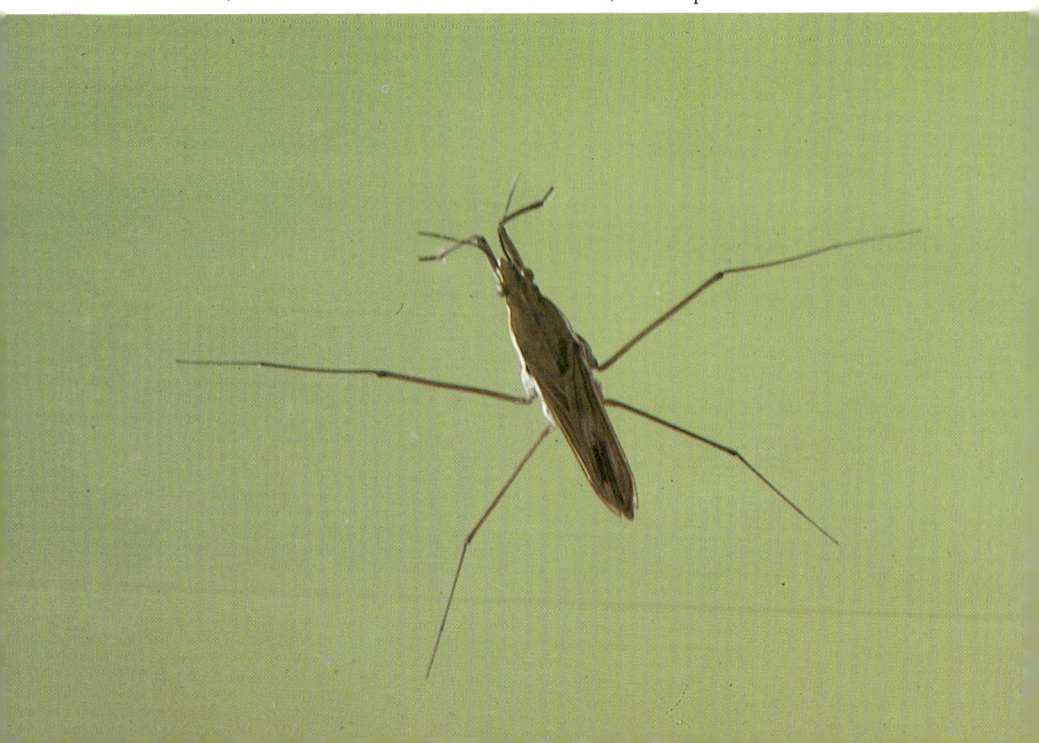

Pond skater, an insect which lives on the water surface, a fierce predator of smaller insects

into one another. In addition to whizzing about on the surface and swimming underwater, whirligig beetles can also crawl about on dry land and occasionally fly.

Mosquitoes, or gnats, lay their eggs on the water and their larvae remain attached to the underside of the surface film until they are ready to shed their skin and fly off as adults. Although mosquito larvae are often seen in ponds, it is easy to observe them by simply leaving a container of water out in your garden during the summer. Tiny, boat-shaped rafts of dark-coloured eggs will almost always appear at the edge of the water. These soon hatch into elongated larvae which hang, head down, from the surface. They breathe through their tails but will quickly let go and wriggle down through the water when you approach too closely.

After a few weeks their shape changes. They appear to have a big 'head' with a small tail curled beneath. They continue to hang beneath the surface, diving down with rapid spinning movements when disturbed. These are the pupae and they do not feed. They are gradually changing inside in preparation for the moment when the adult mosquito will burst out and fly away.

To put two or three larvae in a jam jar of water from the place you captured them must be the easiest way to watch the development of a young insect. They will feed on microscopic plants and animals in the water.

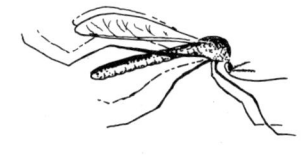

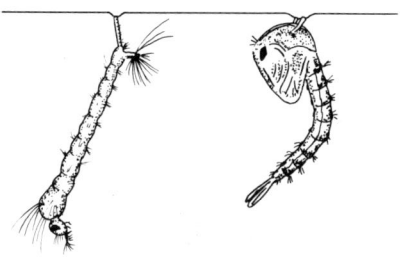

Mosquito: adult (*top*), larva (*left*) and pupa

Cover the jar with muslin so that you can see the adults when they emerge.

Certain mosquitoes commonly transmit malaria to humans in warmer climates, and certain areas of Britain also suffered from this in the past. Nowadays it is extremely rare to catch malaria in Britain. The mosquito which is capable of transmitting the disease breeds in brackish water near the sea.

While the various insects mentioned are the principle inhabitants of this region, there are also various 'visitors'. In particular, snails and flatworms are commonly seen creeping along the underside of the surface-tension skin when the water is calm. Certain spiders also venture out from the water's edge in search of prey.

7
In the mud

The only freshwater environments free from mud are fast streams and the wave-washed margins of lakes, where the movement of the water prevents it from settling. In still and slow-flowing water, the bottom invariably consists of mud, and this is the special home of all sorts of creatures.

What is this mud? Partly, it consists of tiny mineral particles — sand, silt and clay. These are washed into the water from surrounding soil, or are blown in as dust. A large proportion of the mud consists of dead remains of plants and animals in various stages of decomposition. This decomposition does not just happen: it is being brought about by seething masses of bacteria, fungi and microscopic animals. This microscopic life is another vital component of the mud, and includes the microscopic plants known as algae which often give a greenish tinge to the surface layers.

Mud is a rich source of food for those animals able to live in it. Many of them swallow large quantities of it and digest the microscopic animals and plants, as well as some of the rotting dead remains. As in most other environments, these 'grazing' animals are preyed upon by larger carnivores.

The ability to burrow is one typical feature of most creatures found here. Another is the ability to breathe when surrounded by this clogging substance, where there is often a shortage of oxygen. This is a classic example of the way living creatures are always adapted to the environment where they live.

Mud-dwelling animals will invariably be found during the examination of water plants, because a fair amount of mud is pulled up with the roots. But if deliberate examinations are to be made, it is best to scoop up the surface layer of the mud because this is where there is most life. There is more oxygen here and more food in the form of algae (which need light).

The most noticeable inhabitants are the abundant midge larvae or chironomids. Most of them are under 1 cm long, some are bright blood-red, and they swim with an unmistakable lashing, wriggling motion when disturbed. Their worm-like shape is ideal for burrowing and they spend much of their time under the surface of the

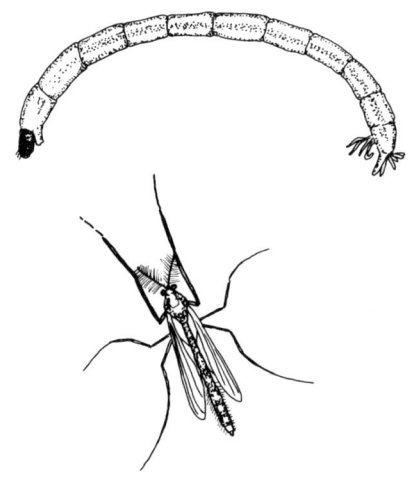

Midge: larva (*top*) and adult

mud. Some species live in silky tubes on the surface of stones or submerged twigs. These tiny tubes, with fine mud adhering to them, can often be seen on the sides of water tanks and goldfish ponds.

Midge larvae feed on organic matter and microscopic life in the mud. They are so abundant that they themselves form a major part of the diet of freshwater fish. The larger red species are the 'bloodworms' often used as bait by coarse fishermen. The red types are most common where the mud is thick and black and oxygen is scarce. Their colour is due to a substance similar to the haemoglobin of our own blood which enables them to absorb enough oxygen even when it is very scarce. Each midge larva eventually turns into a pupa which resembles the adult, with a swollen thorax and traces of folded legs and wings. The pupa swims to the surface, where it sheds its skin and flies away as an adult.

These are non-biting midges. The evil, biting midges are far smaller, less than 2 mm long. They also have aquatic larvae, about 4 mm long and very slender.

True worms may be distinguished from the wormlike midge larvae by their greater length and the absence of the tuft of filaments that a midge larva has at its rear end. They are referred to as 'segmented worms' to distinguish them from the 'roundworms' which may also be seen. (Roundworms have smooth bodies not divided up into segments, and move with a lot of writhing and lashing.)

The earthworm is a familiar segmented worm and many of the aquatic forms look rather like slender miniature earthworms. Masses of thin red tubifex worms (see page 51) may be found on the mud surface, waving their bodies in a concerted effort to increase their oxygen supply. Tubifex worms are a useful food for aquariums.

One group of segmented worms that has given up burrowing and become adapted to a very different way of life is the leeches, no longer true inhabitants of the mud.

Freshwater mussels and cockles are 'bivalve molluscs', relatives of the snails but having two shells. They live on, and in, the mud, ploughing their way through it by means of the single muscular foot which they protrude from between their shells. When the animal is undisturbed, its shells gape apart slightly, and it is possible to see two openings into its fleshy body. Water is constantly pumped in through one of these openings and out of the other. The creature feeds by filtering from the water various microscopic animals and plants swimming and drifting in it, together with any matter stirred up from the mud.

Mayflies have already been described in the section dealing with the inhabitants of the weedbeds. However, certain species of mayfly larvae may also be found in mud samples. These have their front legs flattened for burrowing, but otherwise resemble their cousins amongst the plants, with their three long tails and gill-fringed bodies.

Although comparatively few species are able to live in the mud, those that can meet the special requirements of this habitat are able to multiply profusely, as there is plenty of food there. The mud is seething with life.

Even this invertebrates' paradise has its predators, of course, chief among which is the dragonfly larva. The closely related damsel flies were described in the section on the inhabitants of weedbeds. The dragonfly species vary in form, but generally they are considerably bigger and more sluggish. Another difference is the absence of the three gills at the rear of the abdomen. The dragonfly larva has its own distinctive way of breathing: water is drawn in through the tip of its abdomen, then expelled when the oxygen has been extracted. In this way, the creature is able to produce a jet of water from its backside strong enough to give it jet propulsion.

Any small worm or insect unlucky enough to pass within range of a lurk-

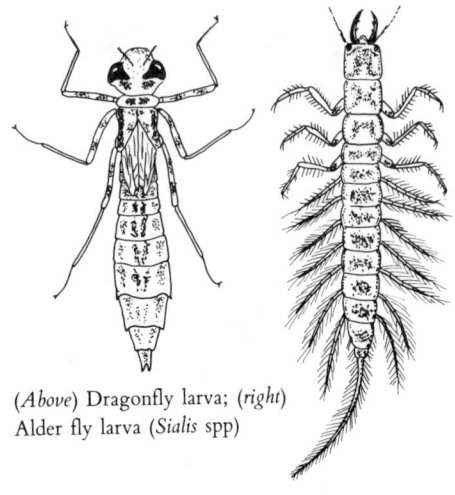

(*Above*) Dragonfly larva; (*right*) Alder fly larva (*Sialis* spp)

ing dragonfly larva disappears in a flash as the extendable jaws whip out and seize it. The beautiful adult dragonfly may be over 10 cm long, depending on the species, and is even more formidable. It is the hawk of the insect world, patrolling a fixed 'beat' of the water margin with a powerful, darting flight, and pouncing on flying insects. A crunching sound can often be heard as it crushes its victim.

The larva of the alder fly also preys on the numerous smaller inhabitants of the mud; its way of life is similar to that of the dragonfly larva. Alder fly larvae can be confusing at first, because they appear to have considerably more than the three pairs of legs expected of all insects. Matters become clearer on close examination, when most of these 'legs' turn out to be long, thin gills fringing the abdomen. (Remember the scarcity of oxygen in this animal's habitat.) The adult alder fly is dark, about an inch long, with transparent, heavily veined wings.

8
Very small animals and plants

Even a small and relatively inexpensive microscope can open the door to the world of tiny freshwater organisms invisible to the naked eye. Such microscopes are often usefully packaged with slides and instruments, intended for use by children. Many just sit on a shelf and gather dust, as the young owner often lacks guidance on what to use it for. It is a different story when the owner of a microscope has an interest in freshwater life. Some of the creatures revealed make the weirdest science-fiction monsters seem tame, while others are objects of breath-taking grace and beauty. Many tiny plants swim around like torpedoes. There is a constant element of surprise when searching through the most unappealing samples of debris brought home from a ditch.

Most junior microscopes are quite adequate. Try before you buy and avoid flimsy, wobbly models, and those which do not focus easily or tend to slip out of focus. Many microscopes have a rotatable turret bearing a selection of objective lenses of varying powers. To have more than one objective lens is not strictly necessary with an inexpensive instrument; in fact the field of view at higher magnifications may well be very limited in a cheap model. High magnification is no earthly use if you can only see a tiny portion of the creature being looked at. In general, a magnification of x50 is perfectly adequate in a microscope lacking expensive lenses. Separate magnifications are normally marked on the objective lens and the eyepiece. Multiplying these two figures together gives the overall magnification of the microscope. Some microscopes have an adjustable mirror to collect light from an electric

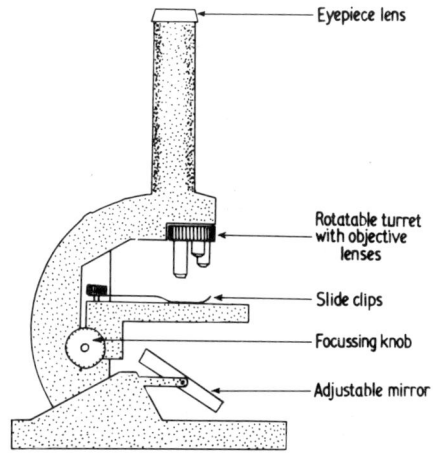

Microscope

bulb or window. Others have a built-in electric light.

A very blurred, distorted image can be produced if you simply dump some wet material on a glass slide for observation. This results from the bending of light, the curved water surface acting as a lens. The solution is to flatten the water on the slide by covering it with an extremely thin glass 'cover slip'. This gives an undistorted image and also reduces the risk of the vulnerable objective lens getting wet. Glass cover slips are extremely fragile and must be cleaned with great care.

To take samples of some algae is a good beginning. These extremely important little plants are, to the unaided eye, perhaps the least appealing of freshwater organisms. Walk slowly along the water's edge, transferring samples to various specimen tubes (or pill containers or small jars) in your pockets. Try to keep each sample separate from the rest. Perhaps into one container will go a waterlogged bit of twig with a haze of greenish-brown slime extending into the water around it. Into another could go a gobbet of bright green slime fished from a green woolly mass covering a patch of mud. Another sample might be taken from the greenish surface of the mud right at the water's edge, and yet another from the bright green, opaque water resting in a cow's hoof-print nearby. Certainly nothing to get excited about so far! Nevertheless, the green or greenish-brown colours were the clues to the presence of algae.

Each sample in turn is prepared for examination. A drop of pondwater is placed on a clean glass slide. A *small* wisp of the sample is placed in this water, using fine forceps or a mounted needle. (A mounted needle can be made by pushing a needle's eye into the eraser end of a pencil.) This tiny sample is then teased out gently across the water drop with two mounted needles.

A cover-slip must be put on to flatten the surface. If this is done carelessly numerous air bubbles trapped underneath will later appear as black circles during examination: avoid this by lowering the cover slip on gently, supporting one end with a mounted needle. If the cover slip now seems to be 'floating', your water drop was too big. If the water has not completely filled the space under the cover slip, the drop was too small. Carefully apply another drop of water to the edge of the cover slip. The flat end of a pencil can be used for this, and the water will quickly be sucked under the cover slip by 'capillarity'.

Set up the microscope with the lowest power (shortest) objective lens, if it has a rotating turret. Looking through the eyepiece, adjust the mirror and/or light source until the illumination is bright and even.

Place the slide on the stage, with the specimen directly under the objective lens. Watching carefully *from the side*, focus the microscope down until the objective lens is very close to the cover slip. If you do not watch from the side, you could crack the slide or, worse, damage the lens.

Now look down the eyepiece and focus slowly upwards until the magnified specimen comes into view. Depending on the sample you took, much of what you now see may consist of grains of sand or silt, magnified like cobblestones, or shapeless masses of decaying plant material. Since the light is coming from below, these will appear dark and opaque. The black circles of trapped bubbles may also be visible. Look out for the green colour of microscopic plants and the movement of microscopic animals. Move the slide slowly with two thumbs as you search.

If you find nothing, waste no more time but make up another slide, perhaps from another of your samples. If you do see something of interest, then you may wish to switch to a more powerful (longer) objective lens by rotating the turret. If the microscope is well made, only a small focussing adjustment will then be necessary. Be careful not to damage the lens against the slide and always focus *upwards* unless watching from the side. Always return to a low magnification when searching for fresh areas of interest on the slide. It makes searching much easier.

If you took your samples carefully as described, looking out for the telltale greenish colour of algae, it will not be long before you are making your first exclamations of wonder at the beauty of these microscopic plants. Perhaps the first you see will be 'filamentous algae'. The many species of this type of algae consist of long, tangled green threads, often

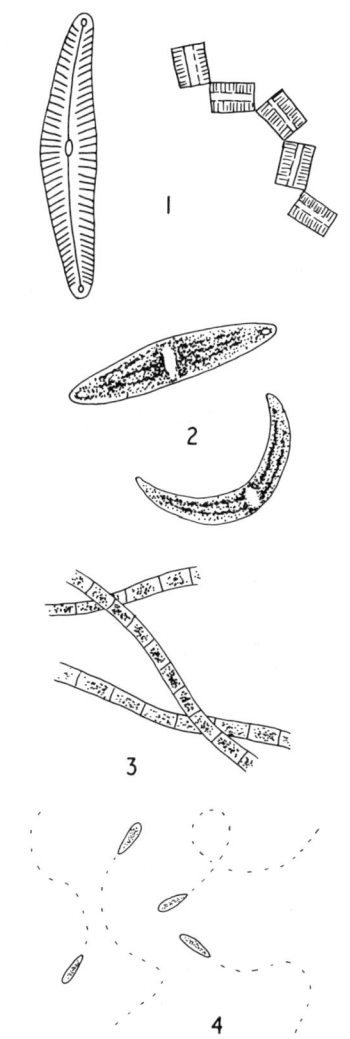

Algae: 1 diatoms, 2 desmids, 3 filamentous algae, 4 flagellates

branching. Under high-power examination each thread is seen to consist of a chain of cells joined end to end, each cell with a complicated internal structure including bright green chloroplasts. Spirogyra is a particu-

35

larly nauseating green slime when handled, but under the microscope it is the most beautiful of the filamentous algae. Its chloroplasts form delicate and lovely spirals within each of its cells.

Small geometrical shapes, coloured green or brown, will sooner or later be seen: discs, boxes, rods, boat-shapes and crescent moons, some in isolation, others joined together in groups and chains. These are diatoms and desmids. They are often seen making mysterious creeping movements which are quite uncanny to watch, sometimes sliding slowly backwards and forwards. Desmids can generally be distinguished from diatoms by their deep green colour and by a very obvious dividing line across their middle.

The swimming algae, or flagellates, have even more startling abilities. They are usually small and first noticed as fast moving green specks, especially in a sample of green water such as that from the waterside hoof-print mentioned before. They are propelled by tiny whips or 'flagella', visible only with a very good microscope. Some species form larger colonies of anything from several to several hundred individual cells, each with its lashing flagella. *Volvox* is the largest of these swimming colonies of microscopic plants, often filling an area of water with pinhead-sized green specks.

There is endless scope for exploring the world of algae, and it can be done at home on the table after a visit to the waterside. At the same time,

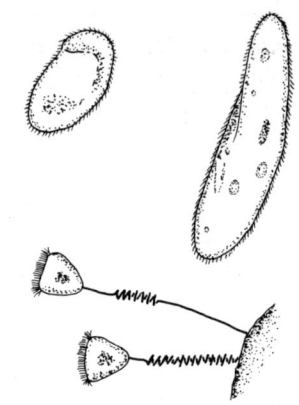

Ciliates

remember the vital importance of these algae. They are the main origin of the 'food chains' in any body of water.

Microscopic animals will always be found, not only in samples of algae but in almost any surface mud or decaying plant material. They differ from the plants in containing no green chlorophyll, and so have to feed rather than manufacture their food like the plants can.

The first microscopic animals you notice are likely to be ciliates. These tend to be small (but not always), colourless and fast moving. Their body surface is covered with numerous short 'whips' called cilia which propel them along. They often rotate as they swim and cross the field of view of the microscope very quickly. If you have sufficient ciliates in a sample, you can often slow them down enough for high-power observation by teasing out a fragment of cotton wool in the water drop before adding the cover slip. Some ciliates are stalked and attached to filamentous algae or other objects. Their pulsating cilia waft particles of food, including algae, towards them.

It is easy to culture ciliates in a 'hay infusion'. Simply boil up some hay or dried grass and pour off the liquid. When this has cooled, add some pondwater and mud and leave for a few days. The water will usually become cloudy as ciliates rapidly reproduce, feeding on the bacteria which are breaking down substances from the hay.

Several species of *Amoeba* may be encountered occasionally. Familiar to anyone who has studied biology at school, this creature is not a ciliate. It moves by 'flowing' rather like an animated blob of glue.

Rotifers might be mistaken for very large ciliates at first glance. They do in fact possess cilia, but these are restricted to the front end and used for feeding. In some species these cilia flicker rapidly on the surface of two projections at the front of the body which consequently look exactly as if they are spinning round. This is the reason for their common name of 'wheel animals'. Many rotifers crawl and swim over plants and mud. Others remain attached by their rear end and a few build themselves protective tubes.

Roundworms, surprisingly, are closely related to rotifers. They are tiny worms, usually much less than half a centimetre long. Their bodies are not divided into segments and they move by vigorous lashing and S-shaped writhing.

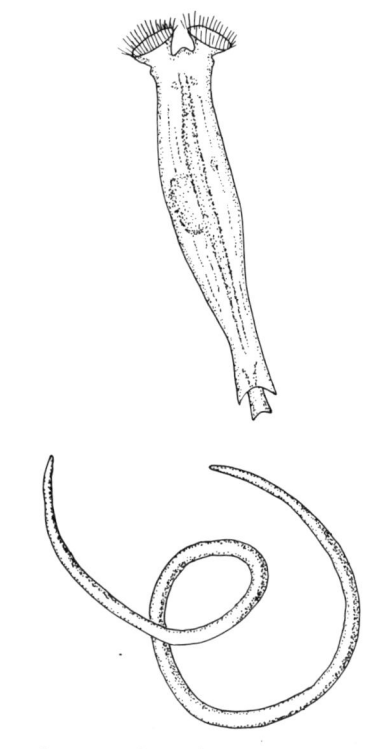

Rotifer (*top*) and roundworm

9
Larger animals

Small mammals are secretive creatures and the thick waterside vegetation often provides ample cover for large numbers of mice, rats, voles and shrews.

Many of these species may also be found in hedgerows (see the *Hedgerow and Wayside* booklet in this series). The water vole, however, is a true inhabitant of the water margin. Sooner or later you will see one if you are quiet, swimming at the head of a V-shaped wake. Often a disturbed water vole will plunge in from the bank and swim rapidly underwater out of harm's way, its unwettable fur glistening with a silver film of air.

It is unfortunate that water voles are sometimes called water rats; in fact they bear little resemblance to rats. They do not have the pointed face of a rat and they are big, heavy looking animals, often up to 20 cm long excluding the tail. They make burrows in the bank, often with an underwater entrance, which explains many a water vole disappearing act; a person may be left scanning the surface and wondering where the vole he alarmed is going to reappear. They feed on plants.

The much smaller water shrew may also be spotted. It is about 8 cm long, excluding the tail, very dark, and has the typical long thin shrew's snout. They are swimming, diving carnivores, feeding on small water animals.

Otters are much less frequently seen, largely because of their intelligence and nocturnal habits. They are not common, but in some areas not as rare as people think. One is more likely to come across their footprints in the mud or even an 'otter slide' down a steep bank where they have been playing.

Waterside birds are a far more familiar sight. Perhaps the moorhen is the most characteristic inhabitant of the marginal undergrowth, and its croaks and squeaks can often be heard when the bird is out of sight. The only similar birds are coots, but these

Water vole (*Arvicola amphibius*)

are larger, have a white rather than a red bill, and prefer the open water of larger lakes.

Various species of duck are also certain to be seen. Wild duck have been hunted for centuries and are consequently extremely shy. Some species dive and swim underwater while feeding, while others, such as the common mallard, feed by up-ending themselves in shallow water. The male 'drakes' are more brightly coloured than the better-camouflaged females. The colours are for courtship purposes and when the drakes moult in the autumn they become virtually indistinguishable from the females.

The great crested grebe is a splendid sight in summer, unmistakable with its large, slender form and two-horned head. The behaviour of a courting pair in the spring is a sight worth seeing, and looks like a primitive, ritualised dance.

Grey wagtail (*Motacilla cinerea*)

Small streams have their own bird-life as well. The grey wagtail is anything but grey, with its bluish back, yellow breast and long black-and-white tail. The dipper is likely to take you by surprise as it streaks past with a shrill 'tweet', following the course of the stream. When it alights on a stone it can be seen to be a small, dark bird with a white breast, bobbing restlessly up and down on its legs. Using the force of the current to hold down its buoyant little body, the dipper walks across the bed of the stream in search of food.

The kingfisher is totally unmistakable and if you are lucky enough to see one you will not quickly forget it. It is very small, all head and beak, but a dazzling orange and blue jewel. It is most likely to be spotted making a rapid flight up or down stream.

The amphibians (frogs, toads and newts) are very closely bound to the water. They have not altered much since the very first four-legged animals crawled out of a prehistoric swamp. When out of water they are restricted to cool, damp places because their thin skin must never dry out. Their ancestors were left behind in the swamps when the first reptiles developed dry, scaly skin and the knack of laying tough eggs on land. So although toads are often found far away from water, they must return to water, along with frogs and newts, to lay their jelly-like strings of eggs every spring.

Newts also venture far from the water after the breeding season is over. They are seldom seen until the following spring, because they are very secretive and emerge to feed mainly at night.

The stickleback is present in most well-established ponds and slow streams. It is quickly recognised from above by its large pair of side fins, the long thin 'wrist' to its tail, and also by a characteristic habit of flexing its body slowly from time to time rather

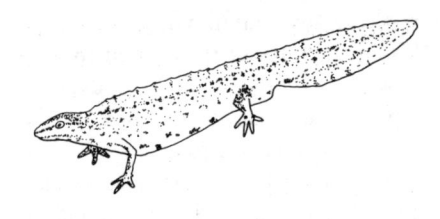

Male smooth newt (*Triturus vulgaris*)

like a cat stretching. Round about May, the males develop their brilliant breeding colours and rival the gaudiest of tropical fish: the back is bluish, the belly a fiery orange and the eyes gleam green as emeralds.

Each male stickleback builds a nest in the spring, using fragments of plant material. Then he coaxes a female to squeeze through the nest and lay her eggs. After he has fertilised them he guards the nest fiercely until the young fish are big enough to fend for themselves.

Sticklebacks will breed freely in captivity, but approximately one square foot of space must be allowed for each male to establish its territory. Even the fat brown female sticklebacks will be driven away before and after the egg-laying stage. It is best to remove females from the aquarium afterwards. Another important requirement to remember is the need for live foods such as waterfleas or chopped worms. Sticklebacks will rarely eat dried fish foods.

Minnows are the fast-water equivalent of the sticklebacks. Shoals of them are found everywhere in rivers, streams and big, stony lakes. When viewed from above, the dark line that runs along the brown body and through the eye is very obvious. They also make attractive and hardy aquarium fish, but obviously require aerated water.

The traditional way to catch minnows is to bash a hole in the funnel-shaped bottom of a clear glass wine bottle, holding the head of a 6 inch nail in place and striking the point with a hammer. Practice makes perfect. The neck is corked and a length of string tied to it. When setting the trap in an area where minnows are shoaling, put in some bread, fill the bottle with water and throw it out from the shore, keeping hold of the string. It is not difficult to catch sixty minnows in one go, so do not take too many. I have found a trap made from a large jam jar even better, using a funnel shaped from perforated-zinc sheeting, held on with elastic bands.

Eels are common in all sorts of freshwater environments. They are all born in the depths of the Sargasso Sea, thousands of miles away. Their leaf-shaped young swim and drift towards the coasts of Europe in the ocean currents known as the Gulf Stream

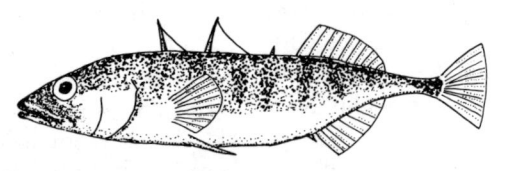

Three-spined stickleback (*Gasterosteus aculeatus*)

and the North Atlantic Current. Their journey takes three years. Many rivers and streams experience an 'elver run' in May, when the young eels arrive at the estuaries and forge their way up stream into freshwater. They now resemble tiny eels about 7–8 cm long and their sheer numbers present quite a spectacle as they struggle up the moss-covered stones at the edges of waterfalls, suffering tremendous casualties from predators. Adult eels too will move overland, especially on wet nights, which explains why they are often found in ponds unconnected with the sea.

Some mature adults return to the sea at the end of the summer, and what happens to them then is unknown.

Other fish might also be encountered in fast streams: speckled young brown trout; the slender, brown loach with its whiskered face; or the ugly bullhead, named after its broad, flat head and enormous mouth.

Most still-water fish will keep well out of your way as you search the shallow margins. But in the spring, you are very likely to encounter the tiny young of the various 'coarse fish' species around weedbeds in shallow water, roach and perch in particular.

10
Aquariums and freshwater communities

The classic beginners' aquarium is the jam jar, which can be a very useful piece of equipment. Generation after generation of the smallest freshwater animals can be cultured in them. Screw-top jars are essential for carrying specimens home and for use as temporary aquariums to examine specimens at the waterside. However, such jars do not make suitable permanent homes for larger animals, as countless sticklebacks and minnows have found to their cost.

It is a shame that so many freshwater animals perish unnecessarily in jam jars, because most of them are easy to keep at home. The usual cause of failure is overcrowding. If the oxygen in the water is used up, the animals suffocate quickly. If their waste chemicals are allowed to accumulate in the water they may be poisoned slowly. So firstly, the number and size of animals stocked must be in proportion to the size of the container. Secondly, a proportion of the water should be changed periodically; how often depends on how crowded it is.

Jam jars *can* be used. It is possible to make a fascinating miniature 'zoo' of pond animals on a shelf or window ledge, consisting of a dozen or so jam jars with explanatory labels. The typical one-pound jar is suitable for small numbers of small animals, for example two or three freshwater shrimps or half a dozen leeches. Two-pound jars are generally better for anything larger. Big glass sweet jars are useful — ask a shopkeeper. The readily available plastic ones tend to contaminate the water and your freshwater animals may not be appreciative of peppermint or aniseed! Tall containers are not really ideal. Those giving the maximum surface area are best for avoiding suffocation. Various items of kitchen ware may be pressed into service, such as mixing bowls. The largest and most active creatures, for example newts or sticklebacks, should not be kept in anything smaller than a plastic washing-up bowl, and a proper aquarium tank is much more attractive.

Let commonsense dictate how many specimens you place in any container. To keep them healthy, try to err generously on the side of caution. You will find out soon enough if you over-step the limit.

Afterwards, keep *well below* that limit. Only clean containers should be used. The word 'clean' needs explaining. A jar that has been lying around in the garden for years, stained with soil and green algae, is probably 'clean' for the purpose of keeping freshwater animals; a bright and sparkling jar taken from the kitchen sink without rinsing off all traces of detergent could well kill its occupants. Jars that still smell of their original contents should be given a prolonged soak.

Half the water in each jar should be replaced every week or so. Larger containers need attending to less often, unless they are heavily stocked. It is useful to have a bucket of pond water standing by for water changes. Tap water can be used if there is no alternative, but it must be allowed to stand for several days first to get rid of any chlorine. Avoid sudden changes in water temperature.

Feeding will be necessary if specimens are to be kept for more than a few days. Remember that it is easy to kill with kindness. Any excess food left in the water uses up oxygen and forms poisons, whether it is live food or dead and decomposing. Keep a close watch and feed no more than is necessary. Small freshwater animals have very diverse feeding habits, and these make an absorbing study. If proper notes are kept of all observations it is possible for a keen amateur to make important discoveries. Animal behaviour has always been a neglected branch of biology.

The following feeding suggestions may be used as a basis for experimenting.

The water flea *Daphnia* is a useful food for many carnivores. It can usually be obtained easily by regular excursions with the plankton net; alternatively, it can be cultured. *Daphnia* feeds on microscopic algae swimming and drifting in midwater (the 'green-water' algae). Algae require sunlight and water rich in nutrients. This can be arranged by keeping a plastic dustbin full of water in the garden with 6 inches of soil in the bottom. *Daphnia* can be added to the water when it has matured enough for algae to develop. Mosquito larvae will also appear and these are another useful food. They have a characteristic way of scattering down from the surface when you approach.

Daphnia may be fed to the following animals: *Hydra*, water mites, water spiders, phantom larvae and those water boatmen which swim on their backs (the 'true' carnivorous water-boatmen).

Daphnia may also be given to some larger carnivores such as water scorpions, dragonfly and damselfly larvae, alder-fly larvae, *some* water beetles and *some* leeches (experiment). But they may prefer something larger. If mosquito larvae are not available, look for a good concentration of freshwater shrimps or any small aquatic insects.

Although some freshwater animals browse on the water plants with stems and leaves, microscopic algae are far more important as food. The animals likely to feed on masses of the woolly 'filamentous' algae, or on the algae coating plants and stones are these — but bear in mind that each category contains many species whose precise requirements vary: snails, caddis larvae, herbivorous beetles (experiment with *Daphnia* in case your beetles are carnivorous), mayfly larvae, freshwater shrimps and water lice.

Many of the leeches which will not attack *Daphnia* or small insects are likely to feed on snails. Those which suck the blood of larger animals like fish will not be convenient to feed, but are likely to live for months without feeding.

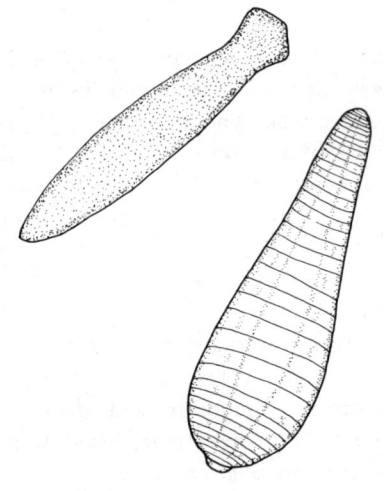

Flatworm (*left*) and leech

Tadpoles and flatworms will feed on small pieces of raw meat and fish. Give very little at a time and remove the surplus. Flatworms are very tolerant of starvation and simply get smaller and smaller. Tadpoles should be released after growing their legs, unless you are confident about being able to supply enough live food for the young adult frogs.

So much for the window-ledge zoo. There is also a less demanding way to keep freshwater life — in a single, balanced aquarium. 'Balanced' means that the various inhabitants, plants, herbivores and carnivores, interact in a more or less natural way so that necessary chemicals in the water are not used up and harmful substances do not accumulate. People who keep fish often talk about a 'balanced' aquarium community consisting of fish, plants and useful micro-organisms, although actually they seldom come close to such an ideal situation, their densely stocked tanks requiring relatively heavy feeding. Much has been made of the fact that plants give off oxygen and use up carbon dioxide produced by animals, so that many people seem to regard their presence as essential for the well-being of aquarium animals. It is forgotten that plants also absorb oxygen and give off carbon dioxide at night, and that dead and damaged plants do this as they decompose. In fact the water surface of the aquarium and any aeration system in use are far more important than plants for the exchange of these gases.

However, it *is* possible to come very close to a balanced, self-sufficient community of pond animals, not requiring much feeding if any — if the aquarium is not overstocked, and if the larger and fiercer carnivores are avoided. Such a set-up can practically look after itself so long as there is plenty of sunlight to support the growth of algae, the primary food of

```
            CARNIVORES
                ↑
            HERBIVORES      Sunlight
                ↑
             PLANTS
```

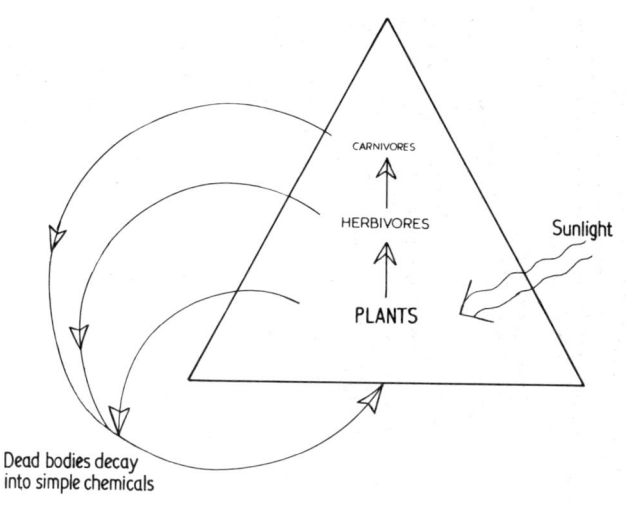

The economics of a freshwater community

(Dead bodies decay into simple chemicals)

any freshwater community. A window-sill is a good location, preferably facing north to avoid direct sunlight.

The aquarium should not be too small; six to twelve gallons is a good size. The unframed glass ones are best, held together with a special adhesive. The small plastic tanks are not all that cheap and easily get scratched. Any aquarium must be placed on a firm, flat base and a sheet of corrugated cardboard or expanded polystyrene will prevent uneven loading and possible cracking of the base.

A quarter of an inch of soil should be spread over the bottom of the tank to help the growth of plants and algae. It should then be covered with an inch of well washed gravel, which is mainly to hold down the roots of water plants and to prevent the soil from clouding the water. A saucer should be placed on top to stop the gravel from being stirred up when the tank is filled. Water from a thriving pond is best because there is no danger of any harmful substances being present. Various suitable species of algae and micro-organisms will also be introduced. Tap water can be used as long as it is left standing for a few days before stocking so that any chlorine can evaporate.

Artificial light will be necessary if the tank does not receive enough daylight. Light bulbs give out a lot of heat so fluorescent strips are better. Algae will be visible after a few weeks. The woolly 'filamentous' algae can be fished out from time to time before the water plants are suffocated. Growths on the front of the tank can be wiped off occasionally, but can be left to grow on the remaining three sides. Dark blue-green algae indicate the presence of too much rotting organic matter — time to change the water completely and clean out the tank.

Cuttings of water plants usually take root quickly and all sorts of small creatures will be noticed in the tank after a few bunches of plants have been pushed into the gravel. Other animals can be added occasionally following collecting trips to the waterside. Some will die, some will survive, many will be eaten. Something like a natural balance will be maintained, provided that the larger predators are avoided.

The regular addition of a wide range of specimens maintains the interest of the display. In such a small 'pond' as your aquarium, many species will inevitably be wiped out completely by predators, leaving you with only a small selection of different species. To compensate for this rule of ecology, inescapable for small habitats, simply dump a wide assortment of specimens into the tank as often as possible. They will sort themselves out remarkably well. The vast populations of tiny micro-organisms in the community will easily take care of any immediate deaths so long as you do not overdo it. In the process, nutrient chemicals will be released into the water that will support further plant growth.

The diagram on page 45 shows the relationship between plants (including algae), herbivores, carnivores and micro-organisms in any balanced environment. It is a relationship which applies not only to freshwater life but to virtually every environment on earth. Notice the 'pyramid' in the diagram. This illustrates another vital rule — that, weight for weight, there is always a greater quantity of herbivores than carnivores in a particular environment, and also a greater quantity of plant material than of herbivores. Compare the amounts of plant material and of plant-eating animals in your aquarium. Also, if the animal community has had time to become balanced, you will see far more plant-eaters than animal-eaters. (Think also of a field: how much grass and other plants, how many mice, how many hawks?)

The reason for this rule is simply that not all the plant material eaten by herbivores is converted into snail, for instance, or caddis larva or water louse. Most of it is 'burnt up' to supply the animals with the energy they need. Similarly, only a fraction of the animal matter eaten by carnivores is converted into beetle, dragonfly larva or stickleback.

It is essential to bear this principle in mind when stocking the aquarium. Always add far more herbivores than carnivores and, above all, make sure there is plenty of light to encourage the growth of algae, on which everything in the freshwater community ultimately depends.

11
Collecting with a purpose

Your first aim when visiting ponds and streams with home-made collecting equipment is to explore as many different habitats as possible. It will be some time before all the creatures mentioned in this book have been found. However, there is a danger that as you become familiar with the commonest species you may cease to find them interesting. This simply means that you have not started to observe their way of life in detail. It is not possible to learn much about an animal simply by glancing at it and smugly repeating its name. There is even a limit to what you can easily find out by consulting books. The 'window-ledge zoo' consisting of rows of jam jars and other containers is the best way to find out more. For example, reference books may describe minute differences in appearance between two closely related species of water beetle; you might well discover for yourself that there are very considerable differences in behaviour, preferred food or in breeding habits. Discoveries of that kind can be made by anyone with a little curiosity.

Once the main inhabitants of your local pond or stream have been identified, it should be possible to work out a 'food-web' diagram, showing how they depend on one another for food.

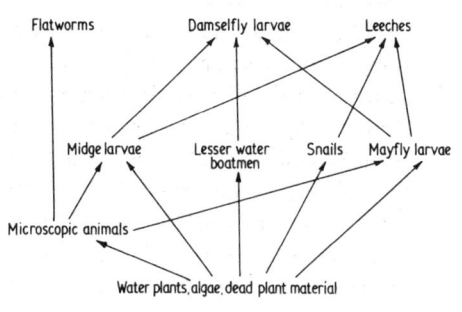

Highly simplified food web for a small ditch community

A detailed knowledge is necessary of what each species is eating. Books may help, but observations on distribution when collecting, and on behaviour in captivity, will be very important. It is easier to make use of such observations if they are written down carefully with all relevant details (and any details that just *might* be important). Observations should be recorded in your notes in an orderly way, not just jotted down at random; a loose-leaf binder is useful. Being scientific is not just a matter of using the correct long words although many scientists do not realise this! It is no more than taking a careful and systematic approach to gathering information. Anyone with a little sense and patience can do that, with or without the scientific vocabulary.

If you have a microscope, never be without a couple of specimen tubes in your pocket. The microscope will continue to reveal surprises long after you are familiar with the commonest of the water creatures that are big enough to see.

After a few collecting trips it will become obvious that the inhabitants of a pond or a stream are not evenly spread out. Different species live in different parts of the habitat. Such information can be set out on a home-made plan of the area being studied. There is much scope for ingenuity when conducting an accurate small-scale survey with such makeshift equipment as a measuring tape, spirit level and lengths of string. Show water depths on the plan, also the position of weed beds (what species?) and any areas shaded by trees (indicate the direction of true north). Other relevant information such as areas trampled by cattle should also be shown.

The next stage is to measure the abundance of various species in each part of the habitat. This has to be done for each of the chosen species in turn. It does not matter if the method used is different for each species, because they are for comparison only. For example you could find out the average number of caddis-larva cases per 10 cm square on the stones along a stream, or estimate the number of great pond snails per square metre in the various parts of a pond, including both weedy and muddy areas. A variety of sampling methods will be needed to make realistic estimates for different species.

One way to present the results would be to make a few copies of the plan, then to write numbers all over each copy indicating the relative abundance of a particular species in different places. Accompanying notes should explain clearly how the numbers were obtained and what they stand for (eg 'Estimated individuals per 10 cm square' or 'Average number of individuals per sweep of handnet after five sweeps').

This sort of study should not extend over more than a week or two because dramatic changes normally happen to populations through the year. These changes are also of great interest and can be studied in various ways. One way is to take a small patch of weed or a single section of a stream, measure the numbers of the commonest species in a given area and repeat this test at intervals through the year; graphs can be drawn showing how their abundance varies month by month.

Alternatively, a single species could be investigated over a year. Evidence of winter migration into deeper water might be found, and it might be possible to follow a yearly life cycle. Young individuals can often be seen growing steadily through the summer, then disappearing in the winter after laying eggs. The average size of animals could be determined at intervals, or the relative numbers of larvae and maturing pupae of such animals as the blackfly might be recorded through the year.

All animals are adapted to their way of life. There are few clearer ways of seeing this principle than by comparing related animals from still and running water. Mayfly larvae are one useful example, those from running water being generally flattened in shape to increase their resistance to being swept away. By studying these and other examples in captivity, many differences can be seen between specimens from such contrasting habitats. Features to examine would be general behaviour, tolerance of low oxygen levels, size of gills, body shape, method of feeding, manner in which adult insect emerges from pupa.

A particularly useful project might be to find out if a suspected source of pollution is in fact causing any measurable harm. Suppose water from a sewage-treatment plant, an industrial estate or a farmyard is draining into a larger stream. Careful estimates should be made of the population densities of various species at points upstream and downstream of the incoming water. Any major differences may indicate that all is not well. The sudden disappearance of stonefly larvae or freshwater shrimps would suggest a drop in oxygen level. A reduction in the numbers of snails and flatworms might be due to poisoning by dissolved heavy metals. A complete absence of animals and plants would probably be caused by rotting organic matter if algae and brownish-grey 'sewage fungus' are present. If even these are absent, some serious kind of water poisoning must be involved. Check the situation carefully upstream of the incoming water, for comparison. The surveys should be done in the warmer months when freshwater life is most abundant. The local Water Authority ought to be interested in any carefully gathered data of this kind.

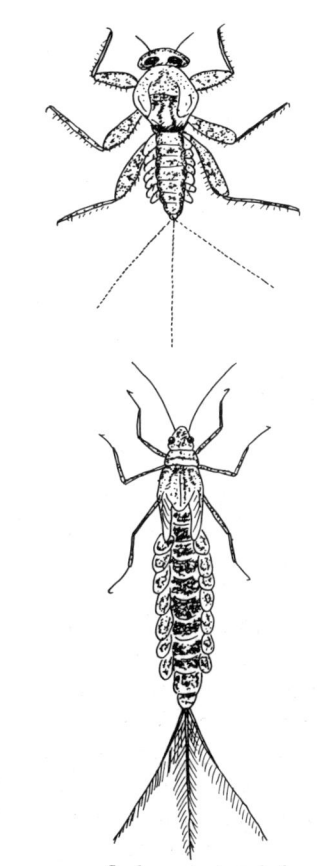

Fast-water mayfly larva (*top*) and slow-water mayfly larva

12
Pollution

There are two ways of looking at water pollution: from the viewpoint of freshwater organisms, or from the viewpoint of man. For example, the drinking water supplied to cities is often heavily dosed with chlorine to kill off any germs or other disease-producing organisms. Few freshwater animals can survive in such water. As far as *they* are concerned, it is polluted.

On the other hand, certain rivers and streams are slightly contaminated with human sewage. Provided the amount does not become too great, this is merely a welcome source of food for many of the inhabitants and does no harm. However, the risk of contracting typhoid fever and other diseases is likely to be high for any human drinking the water. Such water is only polluted from *our* point of view.

Civilised man has lost much of his ancient resistance to disease, and nowadays only clear mountain streams remote from any possible source of sewage are considered fit for drinking, before proper treatment. However there is no health risk at all in studying freshwater life provided that there is not so much sewage in the water that visible signs of pollution are obvious, and that we are not drinking the water. In this section we are more interested in the effects of pollution on freshwater life. Man can take care of himself for the time being.

Sewage *can* harm freshwater life, but in a rather indirect way. The sewage is a rich source of food, especially for all sorts of microscopic creatures. Things get out of hand when too much of it is present. Huge populations of micro-organisms then appear and sometimes use up nearly all the oxygen in the water. It is lack of oxygen that kills off most freshwater animals, not poisoning.

The same effect is produced by any substance that rots — that is to say, any substance that is fed on by micro-organisms. Water draining from an area where a farmer is storing silage is particularly bad in this respect. So are certain types of industrial waste, such as that produced by paper mills.

Streams badly affected in this way are a sorry sight. There are no submerged water plants because these are also suffocated. Slimy masses of algae are the only plants able to survive. Long wisps of greyish-brown 'sewage fungus' on the bottom wave from side to side in the current. These unpleasant growths consist of the various fungi, bacteria and protozoa which feed on the rotting substances and which are tolerant of low oxygen levels.

Some distance downstream from the source of pollution, the organic matter causing the trouble is more or less used up by the organisms feeding on it. Now the oxygen level begins to rise again; the 'sewage fugus' disappears and the first signs of animal life become noticeable again. Tubifex worms, midge larvae and water lice are among the first to be seen. Further still downstream the oxygen level may well have returned to normal with no trace of the pollution remaining.

Even without the use of expensive oxygen-measuring equipment, it is often possible to tell if a stream is slightly polluted. 'Indicator species' such as freshwater shrimps and stonefly larvae need water with a lot of oxygen. If they are absent from streams which otherwise look ideal for them you can begin to suspect pollution of this kind.

Such pollution does not have effects that are permanent, although it may be many years before a cleaned-up habitat is recolonised by the full range of species that previously lived there. Such deoxygenation is basically a *natural* kind of pollution. It tends to occur in small ponds wherever dead leaves accumulate or cattle go to drink, and has been happening for millions of years.

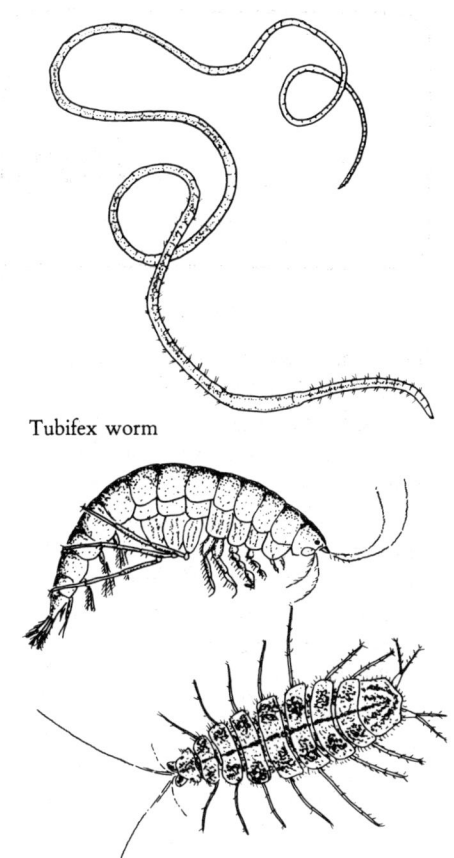

Tubifex worm

Freshwater shrimp (*Gammarus* spp) (*top*) and water louse (*Asellus* spp)

The use of modern chemical fertilisers has resulted in a rather similar kind of pollution, as water drains off farmland. The difference is that these are substances that specifically promote the growth of plants, rather than 'sewage fungus'. Heavy growths of fast-growing algae may choke other plants or turn the water green. A lethal fall in oxygen level is likely when the algae dies and rots at the end of the summer or for any other reason.

Apart from such 'indirect' pollution, there are many substances which do actually poison water creatures. Such pollution is widespread, but often more selective than the kind which removes oxygen. For example, the absence of flatworms and snails from certain streams near old mine workings is the result of poisoning by small quantities of dissolved heavy metals. Insects seem to be less affected by this. Although only part of the freshwater community is affected, this sort of pollution is just as disturbing as deoxygenation because heavy metals are normally removed from the water by various organisms at only a very slow rate. Worse, they are 'cumulative poisons' which gradually build up in the bodies of such creatures as water birds before their eventual death. An affected animal will also pass on the poison to any other animal that eats it.

Old mine workings are not the only source of dissolved heavy metals; these are also present in the industrial wastes that enter our rivers. Many other poisons too are produced by industry, and again many of them are cumulative poisons that can be passed on to predators eating affected animals. To the small ponds and streams that we are considering, the greatest danger of this form of poisoning comes from the insecticides farmers spread on their crops. These tend to be washed into the nearest stream and many of them do not decay but continue to harm populations of living creatures in ways that are often hard to study.

Other substances entering the water harm the inhabitants but not by poisoning. Water draining from certain mines and quarries may contain so many fine suspended particles that plants cannot get enough light to grow. In this way the primary source of food can be cut off, resulting in a barren habitat. The same particles may also clog the gills of some creatures, or completely smother and bury others. The same results can be caused by trampling cattle in a small pond or by boats in a shallow boating lake, stirring up the mud.

I recall a boating lake in a Liverpool park that developed crystal-clear water and a luxuriant growth of water plants when the rowing boats were taken out for overhaul one summer. By the time the boats came back the plant roots had stabilised the mud, preventing the boats from stirring it up, so the water stayed clear and the plants kept on growing. When the weedbeds reached the surface, however, the park authorities received complaints from anglers and treated the water with a fairly harmless weedkiller. The plants rotted quickly in the warm weather, using up a lot of oxygen in the process. Many fish died. A few weeks later the boats were stirring up the mud again and everything was murky normality.

It is a wonder that freshwater life still survives. Yet it certainly does: indeed, it is flourishing and any new, man-made habitats are colonised with astonishing speed.